YOUR KNOWLEDGE HAS VALUE

Cosmas Alfred Butele

A Review of Honeybee Biodiversity, Behaviour and Management

GRIN Verlag

Bibliografische Information der Deutschen Nationalbibliothek:

Die Deutsche Bibliothek verzeichnet diese Publikation in der Deutschen National-
bibliografie; detaillierte bibliografische Daten sind im Internet über http://dnb.d-
nb.de/ abrufbar.

Imprint:

Copyright © 2012 GRIN Verlag GmbH
Druck und Bindung: Books on Demand GmbH, Norderstedt Germany
ISBN: 978-3-656-37413-8

This book at GRIN:

http://www.grin.com/en/e-book/209165/a-review-of-honeybee-biodiversity-beha-
viour-and-management

GRIN - Your knowledge has value

Der GRIN Verlag publiziert seit 1998 wissenschaftliche Arbeiten von Studenten, Hochschullehrern und anderen Akademikern als eBook und gedrucktes Buch. Die Verlagswebsite www.grin.com ist die ideale Plattform zur Veröffentlichung von Hausarbeiten, Abschlussarbeiten, wissenschaftlichen Aufsätzen, Dissertationen und Fachbüchern.

Visit us on the internet:

http://www.grin.com/

http://www.facebook.com/grincom

http://www.twitter.com/grin_com

BUTELE COSMAS ALFRED

COURSE TITLE: APIOLOGY AND APICULTURE

A REVIEW OF HONEY BEE BIODIVERSITY, BEHAVIOUR AND MANAGEMENT

ATLANTIC INTERNATIONAL UNIVERSITY
HONULULU, HAWAII

© 11/06/2012

TABLE OF CONTENTS:

LIST OF TABLES:

LIST OF ACRONYMS:

μg	Milligrams
ABP	Acute Bee Paralysis
ABPV	Acute Bee Paralysis Virus
ABPV/APV	Acute bee paralysis virus
ABV	Arkansas Bee Virus
AFB	American foulbrood
B.C.	Before Christ
B/D	Bees *for* Development
BQC	Black Queen Cell
BQCV	Black Queen Cell Virus
CBP	Chronic bee paralysis
CBPV	Chronic bee paralysis virus
CCD	Colony Collapse Disorder
CWV	Cloudy Wing Virus
DWV	Deformed Wing Virus
e.g.	for example
EC	Emulsifiable Concentrate
EFB	European foulbrood
et al.	and other people
GWM	Greater Wax Moth
IAPV	Israel acute paralysis virus
IBRA	International Bee Research Association

KBV	Kashmir Bee Virus
KTB	Kenyan Top Bar
LHB	Large Hive Beetle
LWM	Lesser Wax Moth
MAAIF	Ministry of Agriculture, Animal Industry and Fisheries
NaLIRRI	National Livestock Resources Research Institute
NaOH	Sodium Hydroxide
NARO	National Agricultural Research Institute
PDB	Paradichlorobenzene
SBV	Sacbrood virus
SHB	Small Hive Beetle
UK	United Kingdom
US	United States
USSR	Union of Soviet Socialist Republics

Introduction

Bees evolved in specific areas of the world long ago, before they spread to become globally as they are today (Tables 1 and II), according to Kugonza (2009). As they spread, they became adapted to the local ecological conditions of the different areas, changing in morphology and behaviour to fit within the requirements of the ecosystem, giving rise to a wide bee biodiversity of bee species and races we see today. Bees are classified under Animal *Kingdom, Phylum* Arthropoda. They belong to *Class* Insecta, which is divided into 29 *Orders*. Bees belong to the *order* Hymenoptera, which has three *Super families,* namely: Apoidea (bees), Formicoidea (ants) and Vespoidea (wasps). There are around 30,000 named species of bees (Apoidea). Apoidea is further divided into several *Families,* namely Apidae (social bees), colletidae, Andrenidae, Halictidae, Melittidae, Megachilidae, and Anthophoridae. Most of the *Families* have solitary individuals: each female bee makes her own nest, lays a single egg and provides food for the single larva that develops. However, a high level of social development is shown by the species in Apidae where the individuals live together in a permanent, large colony, headed by a single egg-laying queen (B*f*D, 2003c). Apidae is composed of four *genera:* Apis (honeybees), *Trigona* and *Melipona* (stingless bees), and *Bombus* (bumble bees). According to MAAIF (2012a), the *genus Apis* is comprised of 5 main species of honey bees: *Apis dorsata* (the giant honey bee); *Apis laboriosa* (the darker giant honey bee); *Apis florea* (the little or dwarf honey bee)*; Apis cerana* (formerly *Apis indica),* is the eastern hive honey bee*;* and *Apis mellifera* (western honey bee). These species have evolved and differentiated into more species and races of honey bees. Kugonza (2009) described 9 species of honey bees in the world, which Oldroyd and Wongsiri (2006) grouped under 3 subgenera: Micrapis (*Apis florea* and *A. andreniformis),* Megapis (*A. dorsata* and *A. laboriosa) and* Apis (*A. cerana,* recently recognized as separate races of *A. nigrocinta, A. koschevnikovi* and *A. nuluensis,* and *A. mellifera).* Dietz (1992), Hussein (2000) and Wikipedia (2012) described over 28 races of *A. mellifera* alone. Beekeeping started with honey bees (*Apis* species), a practice called Apiculture, although keeping of stingless bees, belonging to the genera *Trigona* and *Melipona*, a practice called Meliponiculture, has recently picked up, as reported by B*f*D (2003a), Fajardo and Cervancia (2003) and Braga *et al.,* (2009). The main focus of this paper is Apiculture. Evidence of rock paintings on mountain shelters in Spain point that honey bee (*Apis*) hunting existed as early as 7000 B.C. (Mesolithic times). With time and increasing human population and pressure, the honey bee resource started disappearing. To reduce the hardship and unpredictability of harvesting from wild honey bee colonies, people found ways to increase ownership and management of honey bee colonies kept in hives. The earliest hive probably was a log from a fallen tree in which wild honey bees had nested. Cork and other types of tree bark were then used for hive making and later on, hives were made from planks cut from tree trunks. The first centres of honey bee culture were in the Middle East, in dry and open country (most likely Tropical India and Southeast Asia). The first recognized hives were pottery vessels made during the Neolithic period, from about 5000 B.C. onwards. Woven baskets used as hives came much later on. At this stage protection was provided to the honey bee colony in return for periodic harvests of honey. The idea was to maintain the colony for future harvests instead of destroying it for a one-time harvest. Little was understood as to what was

going on inside the hive since events could not be seen. It was not also realized that the large bee was actually a female, the queen bee, which is the mother of all the bees in the hive. It was mistakenly referred to as the "King bee". The sexes of the workers and drones and the facts of mating between the queen and drones were not known. Neither did man know about bees themselves secreting wax used to build the comb. The relationship between bees, flowers and formation of seeds and fruits was also not known (Kugonza, 2009; and MAAIF, 2012a). Studying the behaviour of bees inside their nest in dark cavities with small entrances was not possible until the development of the glass-walled observation hive about 200 years ago (Gary, 1992). This advancement helped to understand the lifecycle of bees, bee biology, activities and behaviour of bees inside the hive. However, still, no suitable hive was found until 1891 when Lorenzo Lorraine Langstroth, an American Church minister, discovered the concept of bee space and made the first commercial bee hive, which to-date bears his name. Langstroth is now accepted as the father of modern beekeeping. Today, man is able to exploit the honey bee resource, to a certain degree of sustainability, for commercial pollination services, honey, beeswax, pollen, propolis and bee venom. Clearly from the foregoing explanations, so far, advancements in beekeeping have hinged on man's understanding of the activities and behaviour of bees. Although, Kenyan Top Bar (KTB) hive, a movable-comb hive technology, has been invented alongside the Langstroth (movable-frame) hive technology to suit tropical African bees, it has not worked well. The bees are still less settled, more nomadic or migratory, annual, more aggressive/defensive and sting prone, with higher absconding and swarming rates, and are therefore more difficult to manage than their temperate counter parts (Paterson, 1999; and Rinderer, 1999). This paper therefore reviews the various honey bee species and races and the characteristic behavior of each species and race, the factors affecting the behavior, and how this knowledge in totality is applied to better manage the honey bee resource.

Description

Honey bee activities and behaviour can be categorized into 2; those displayed under normal conditions and the others displayed under conditions of stress or disturbance. In both cases, the bees are responding to factors or stimuli in their immediate internal and external environments, which are detected by their sensory cells. The bees react to the stimuli in a particular (stereo-typed) way, because their nervous systems are "hard-wired" or "programmed" genetically to react in this manner (Gary, 1992), hence their behaviour. This section, therefore, describes the internal and external stimuli that bees respond to, the pattern of their responses and how the behavioral responses can be manipulated for the benefit of mankind. The various honey bee species and races, their characteristic behaviors and geographical distribution are also described here (Tables 1 and II).
1.   The internal factors affecting honey bee activities and behaviour include:-
(i)   The caste and sex of the bee: These determine what activities a particular bee will carry out in the colony. For example, the queen is a female with well developed ovaries and the only fertile female in the hive. Its main function is to lay eggs. The only function of the drone bee, which is male, is to mate with the queen. The role of the queen bee and the drone bee is, therefore, to reproduce the species. In addition to the reproductive role, the queen bee controls the activities of all the

bees in the colony through pheromone communication. While, worker bees are sterile females and perform a number of activities in the colony including cell/house cleaning; brood, queen and drone feeding; wax secretion and comb building; food transmission and processing; colony defence; undertaking; thermoregulation; foraging; and robbing (Kugonza, 2009; and MAAIF, 2012a). This knowledge is very crucial for the beekeeper to understand which caste is not performing its duty and why, and then design appropriate remedial actions to be taken.

(ii)   Stage of development: The activities of worker bees vary with age of development, a phenomenon described by Gary (1992) as "temporal division of labour" or "age polyethism". For example, within a day or so after emergence, young worker bees begin to feed nectar, diluted honey and pollen to larvae more than 3 days old. During the first 3 days after emergence, they typically clean the hives from which bees have recently emerged. At approximately 6 to 12 days of age, after their brood food (hypopharyngeal) glands are mature enough to secret royal jelly, they begin to feed young larvae less than three days old. Wax secretion starts until the glands are mature enough. Furthermore, stinging behaviour starts when the stinging structure has developed fully. Also, flight is impossible for very young bees because their nervous system and muscles are not fully developed. This aspect of knowledge can help in designing colony multiplication, queen rearing, royal jelly and bee venom production projects.

(iii)   Hormones: The hormone system of bees consists of two hormones. The juvenile hormone, in the larva, is responsible for slowing down the rate of development from the stage of pupa to adult. The ecdysone (moulting) hormone is an antagonist to the juvenile hormone and is responsible for quickening the process of development. The ratio between the hormones determines the duration of development (Kugonza, 2009). These hormones may be of use to bee breeders. They may also be of use in bee pest and disease management packaging to delay or quicken the development of brood in relation to the period of occurrence of brood pests or diseases.

(iv)   **Physical stimuli** inside the body of a bee also affect behaviour. For example, the sensing of the stretching of the honey stomach either stimulates or inhibits feeding (Gary, 1992). The knowledge of this fact helps one to determine and plan diurnal feeding requirements of his or her honey bee colonies.

(v)   Genetic composition: The genetic composition of the bee is now known to exert a major effect on behaviour. Each bee tends to express different behaviors according to its genetic profile. For example, Rothenbuhler (1968) in Gary (1992) demonstrated that 2 specific genes control the house cleaning behaviour of worker bees. Worker bees endowed with one of the genes uncapped the cells, but did not remove the dead brood. Other workers that contained the other gene did not uncap cells, but removed dead larvae or pupae from cells that had been uncapped. Enough workers with both genes must be present in the colony for effective house cleaning activity. This has a bearing on disease resistance, especially the American foulbrood (Gary, 1992). Genetically controlled behavioral characters of bees have a wider application in selection for desirable qualities and breeding, and biotechnology.

(vi)   Bee activities frequently are regulated by internal, physiological "time clocks" that trigger specific behaviors at specific times, especially on the 24 hour cycle rhythm. Bees, therefore, remember the time of day and tend to arrive at nectar and pollen sources at the correct time (Gary, 1992) or to cease foraging activities especially at night when there is complete darkness (Paliwal *et al*. 2005). This information helps to design appropriate confinement time table for bees during a pesticide spraying programme in an area, as reported by Rashad, *et al*. (1985) or to schedule harvesting at night.

(vii)   The species and race of bees makes them behave differently characteristic of that species or race. Specific management packages can be developed for each species or race based on its activities and behavior.

2.   The external factors affecting bee activities and behaviour are:-

(i)   **Sounds, chemicals or odors, touch, light and magnetic fields** are detected by thousands of specialized sensory cells. Nerve impulses from these cells speed along neural pathways of the nervous system and cause the bee to behave in a stereotyped manner when stimulated by the appropriate mixture or "configuration" of stimuli. This fact has led to the discovery of the pheromone communication system in bees, and the development of lures for bee management (Ferguson, 1985), although Gary (1992) deviated from the principle slightly, saying individual bees do not always respond the same way to identical stimuli, owing to their differences in sensitivity.

(ii)   Climatic and ecological factors: Bees in regions with generally low temperatures and reliable rainfall pattern, especially in temperate climates, tend to be gentle and perennial, making their management easy. While in the tropical Africa, where there are generally high temperatures, the bees are nomadic or migratory, annual, more defensive and prone to sting, making their management difficult up to today. The available packages for tropical African bee management are imported technologies, were developed based on the European honey bee behaviour, and are therefore not satisfactory. It is the very reason for this study to find out more about honey bee behavior, particularly the African honey bees! Bee behavior also varies with specific adaptation to different ecosystems. For example, there are 3 subspecies of *Apis mellifera* which inhabit the East African ecological zones. *A. m. litorea* occupies the coastal plain below 500 metres above sea level. Above the coastal escarpment, the major subspecies, *A. m. scutellata*, is found. This species ranges throughout most of East Africa. However, where mountains cause rain shadows, that rainfall stability has caused the evolution of a third species, *A. m. monticola*. This subspecies is found between 2,000 and 3,000 metres above sea level. In areas where the ranges of these 3 subspecies meet, hybrid zones exist where the honey bees have intermediate characteristics. Hence the 3 species form a graded, stepwise ecocline along the altitudinal range. *A. m. monticola* is considered to be more perennial, gentle and a very good honey producer, *A. m. scutellata* is migratory, prone to abscond and sting, and in comparison, a poorer honey producer. *A. m. litorea* is similar to *A. m. scutellata*, except it is more prone to sting and perhaps less migratory due to living in a range with somewhat more predictable rain. East African beekeeping ecological zones with their different subspecies are a useful genetic preserve. *A. m. scutellata-A. m. monticola* hybrid

zone may provide breeding stock that better suits beekeepers and is still adapted to local ecology (Ruttner and Kauhausen, 1985; Rinderer, 1999). Overall, there is a wide bee biodiversity in the world (Tables 1 and II) because of specific adaptation to different climatic and ecological factors.

(iii)   Diseases, parasites, pests and predators: High loads of pests, predators, parasites and diseases on bee colonies interfere with the normal functions of the bees, hence affecting their behavior. Viruses, fungi, protozoa and bacteria are known to infect bees (Table 3). The infection interrupts the normal behavior and activities of the bees, sometimes destroying the entire colony, hence considerable losses to the beekeeping world. Bees are also attacked by a number of pests and predators, including man, wax moth, honey badger, mites (e.g. *Varroa jacobsoni*), etc. (Table IV). The mites are also known to vector bee viruses, further compounding the situation (Akratanakul, 1985; Clauss, 1985; Corner, 1985; Mwangi, 1985; Xianshu, 1985; Woyke, 1985; Ingemar, 1999; Kugonza, 2009; MAAIF, 2012a).

(iv)   Chemical poisoning: Various agricultural chemicals or pesticides are used to control plant pests. Atkins (1992) gave the following examples of the pesticides used: acaricides, antibiotics, chemosterilants, defoliants, dessicants, fungicides, herbicides, insect growth regulators, insecticides, nematocides and plant growth regulators. Unfortunately, the honey bee is susceptible to many of these pesticides used. As a result, the honey bee is subjected to the hazard of chemical poisoning, which may be the most serious problem in beekeeping. When poisoned the worker bees lose their sense of orientation and/or their ability to fly and usually die, hence leading to low productivity and death of colonies. The first sign of pesticide poisoning is the appearance of large numbers of dead or dying bees at the colony entrances throughout the apiary. Examples of pesticides toxic to bees are summarized in Tables 5, VI and 7. This knowledge helps the bee keeper to carefully select the correct and recommended pesticides (relatively nontoxic to bees) for use in the area or to make bio pesticides that do not kill bees, to schedule pesticide application programmes at such times when the worker bees are not out in the field actively foraging, night time preferably or to confine the bees when spraying, and to educate the neighboring crop and livestock farmers accordingly (Rashad, *et al*. 1985; MAAIF, 2012a).

Table 1: A summary of the races of the Western honey bee, *Apis mellifera*, their Behavioral and Morphological Description, and Geographical Distribution:

S/No.	Races of *Apis mellifera*	Behavioral and Morphological Description	Geographical Distribution
A. African Group			
1.	*Apis mellifera scutellata* Lepeletier (1836)	It is referred to as the African bee; the most defensive bee race, can attack a human or an animal with 500 to 5000 stings; a small bee with one or two yellow bands on the abdomen and a bright yellow scutellum on the thorax; nests in cavities in trees, anthills, the ground and houses; high rate of swarming, migration or nomadism and absconding; and in comparison with other races, a poorer honey producer.	Kenya, Uganda, Tanzania, Ethiopia, South Africa and Central Africa

2.	*A. m. adansonii* Latreille (1804)	It is referred to as the west coast bee. It used to be called the common African honey bee. It is also aggressive/defensive; prone to swarm, migrate and abscond.	West African coast, from Senegal to Congo, including Mali, Benin, Guinea Bissau, Burkina Faso, Ivory Coast (Cote d'ivoire), Nigeria, Togo, Sierra Leone and Ghana
3.	*A. m. litorea* Smith (1961)	Similar to *A. m. scutellata*, except it is more prone to sting and perhaps less migratory due to living in a range with somewhat more predictable rain.	East coast of Africa, from Somalia to Mozambique including Kenya
4.	*A. m. monticola* Smith (1961)	It is referred to as the black mountain bee; perennial and very gentle; a very good honey producer; nests in the cool rain forests of mountains at high altitudes; able to maintain its racial integrity even though perennial hybridization takes place with *A. m. scutellata* in a fluctuating transitory zone	Rain forests of the East African mountains (Ethiopia, Kenya and Tanzania)
5.	*A. m. lamarckii* Cockerell (1906)	It is referred to as the Egyptian bee	Egypt, Somalia and Sudan
6.	*A. m. capensis* Escholtz (1821)	It is referred to as the Cape bee; high rate of female parthenogenesis of worker bees	Southwest coast of South Africa
7.	*A. m. unicolor* Latreille (1804)	The darkest honey bee; island honey bee race of Madagascar	Madagascar
8.	*A. m. yemenitica* Ruttner (1975b)	Inhabitant of arid zones	Kenya, Sudan, Chad, Ethiopia, Eritrea, Somalia and areas of Africa opposite Yemen
B. Central Mediterranean and Southeastern European Group			
9.	*A. m. sicula* Montagano (1911)	Sicilian honey bee; dark in color; the hair of the thorax of workers and drones is yellowish; quiet and gentle during manipulations; brood rearing occurs almost year round; resident queen and daughter queens can live together peacefully in the nest prior to departure of the swarm	Sicilia
10.	*A. m. ligustica* Spinola (1806)	It is also called the Italian honey bee; usually yellow but can be "leather-like"; most popular race worldwide; adapted to mild, wet winter of	Italy, Germany, US, Finland

A Review of Honey Bee Biodiversity, Behaviour and Management

6

		Mediterranean and therefore continues well into the English winter; requires large winter structures because of large colony size; starts laying heavily in late winter/early spring; low spring build up and requires heavy feeding; easily congested and so swarms; very docile; little use of propolis; drifts; a robber; requires large to very large hive; excellent forager and good comb builder.	
11.	*A. m. carnica* Pollmann (1879)	It is called the Carniolan honey bee; steel grey in colour and slender; very quiet and gentle (exceptional docility); very prone to swarming; rapid spring build up; and requires medium sized hive; has beneficial characteristics of longevity, hardiness, foraging ability and wintering ability; but not yet excellent honey producer and comb builder; not a robber.	Switzerland, Upper Carniola, Carinthia, Styria, Yugoslavia, Rumania, Bulgaria, Hungary, Austria, Germany, England, US
12.	*A. m. macedonica* Ruttner (1988a)	Macedonian honey bee; very gentle; disinclination to swarm; strong winter population	Macedonia, England
13.	*A. m. cecropia* Kiesenwetter (1860)	Greek honey bee; varied colors; excellent forager; strong colonies in terms of population; good temperament; disinclination to swarm; gentle and quiet during examination; tendency to propolize; construction of brace-combs; brood rearing is restricted to a particular month of the year	Greece
C. Near East Group			
14.	*A. m. anatoliaca* Maa (1953)	The central Anatolian honey bee; small in size with a smudgy orange color which becomes brown on the posterior dorsal and ventral segments; scutellum is generally dark orange in color; disposition to build excessive brace-combs; and excessive use of propolis; highly susceptible to low temperatures	Anatolia, Turkey
15.	*A. m. adamii* Ruttner (1975d)	Highly defensive when kept in cool climate; winters without difficulty in a cold-temperate climate; brood rearing activity continues during winter	Island of Crete, England
16.	*A. m. caucasica* Gorbachev (1916)	Gray Caucasian honey bee; dark to black in colour with brown spots on the first bands of the abdomen; hairs of the worker are short and gray; thoracic hairs of drones are black; longest tongue of any economically important race of honey bees; gentle and calm on the comb; can forage at lower temperatures and less favorable climatic conditions; does not drift to other hives and winters well; large use of propolis; construction of brace-combs	Central Caucasus high valleys, Turkey, Western Europe, US and Russia

D. Western Mediterranean and Northwestern European Group			
17.	*A. m. sahariensis* Baldensperger (1922)	Inhabits the western Sahara zone; linked to the Central African honey bee, *A. m. adansonii*, and the western and northern European honey bees, *A. m. iberica* and *A. m. mellifera*	Western Sahara and Morocco
18.	*A. m. intermissa* Buttel-Reepen (1906)	Inhabits the northern Sahara zone; linked to the Central African honey bee, *A. m. adansonii*, and the western and northern European honey bees, *A. m. iberica* and *A. m. mellifera*	Morocco, Tunisia, Algeria and Libya
19.	*A. m. iberica* Goetze (1964)	Called the dark bee of the Iberian Peninsula, said to be the link between the African and the northwestern European groups; appears very closely like the *A. m. mellifera*, the dark bees; quick defence reaction; nervousness on comb; heavy use of propolis; and propensity to swarm.	Iberia, Argentina, Spain and Portugal
20.	*A. m. mellifera* Linnaeus (1758)	Is dark brown, almost black with a shiny large, burly body; some non-selected strains can be defensive; very winter hardy; goes readily into winter with minimum preparation; minimum winter food stores due to small cluster; easily stimulated to produce strong foraging colony in April; starts early in the morning and finishes late; flies at 6°C; flies in winds; flies in light rain; smallish colonies; does not swarm easily; suitable for the whole of UK; requires small to medium hives; it is the ideal hobbyist bee-requiring just the basic management.	Ireland, US, France, British Islands, Scotland, Central Europe north of the Alps, Northeastern Europe, the plains of North Poland, USSR and Norway

Sources: Crane, 1985; Mbaya, 1985; Ruttner and Kauhausen, 1985; Dietz, 1992; Rinderer, 1999; Hussein, 2000; BfD, 2003a; BfD, 2003b; BfD, 2006; Clare, 2007; Kugonza, 2009; Nombre, *et al.*, 2009; and MAAIF, 2012a.

Table II: A summary of the species and races of Oriental honey bees, their Behavioral and Morphological Description, and Geographical Distribution

S/No.	Species and Races of Oriental honey bees	Behavioral and Morphological Description	Geographical Distribution
1.	*Apis dorsata* Fabricius (1798)	It is referred to as the rock bee; lives in the open on a large single honey comb hung from the thick branch of a tree, a rocky cliff or the eaves of a building; the comb measures 2-3 metres across and usually located 3-20 metres from the ground; and the workers are 17-19 mm long; huge deposits of honey; high propensity to swarm; and high ferociousness.	Malaysia, Borneo, India, Philippines, Vietnam, Bangladesh, Sri Lanka and other South-East Asian countries
2.	*Apis laboriosa*	It is referred to as the darker giant bee, a specialized mountain or rock bee; it is actually the largest bee; high propensity to swarm	Nepal, Bhutan, Malaysia, India, Borneo, Thailand and other southeast Asian countries

A Review of Honey Bee Biodiversity, Behaviour and Management

3.	*Apis florea* Fabricius (1787)	It is referred to as the little or dwarf bee; about 10 mm long and about half the size of *A. mellifera*; similar in appearance to *A. andreniformis* but more reddish; is open-nesting and construct a single comb on a tree branch, like *A. andreniformis*; the straight run portions of the wag-tail dance point directly to the food source, as the dance is done on the horizontal projections of the comb.	Oman, India, Thailand, Philippines and other Southeast Asian countries
4.	*Apis andreniformis*	Called the small dwarf honey bee of south east Asia; same size as *A. florea*; similar in appearance to *A. florea* but darker and the first abdominal segment is totally black in old bees; is open-nesting and construct a single comb on a tree branch, like *A. florea*	Southern China Peninsula, Malaysia, Borneo
4.	*Apis cerana*	It is referred to as the eastern hive honey bee; has nests that consist of several parallel, vertical combs; dark cavities usually used as nesting sites; similar to *A. mellifera* and can be kept in *A. mellifera* hives, but produces smaller colonies; has tendency to abscond	Central and Northern China, Southern Russia, India, Japan, Malaysia, Nepal, Thailand, Philippines, Afghanistan, Borneo, and other Asian countries
5.	*A. koschevnikovi* Buttel-Reepen (1906)	Cavity dwelling; occupies multiple comb nests; similar to *A. cerana* but with much larger cubital index; weak defensive behavior	Northern Borneo, Malaysia
6.	*A. nigrocinta.*	Cavity dwelling; occupies multiple comb nests; a subspecies of *A. cerana*	Thailand, Philippines, Afghanistan, Borneo, and other Asian countries
7.	*A. nuluensis*	Cavity dwelling; occupies multiple comb nests; a subspecies of *A. cerana*	Thailand, Philippines, Afghanistan, Borneo, and other Asian countries

Sources: Crane, 1985; Dietz, 1992; Rinderer, 1999; Rai, 1999; Ahmad, *et al.*, 2003; B*f*D, 2003b; B*f*D, 2003c; Paliwal, *et al.*, 2005; Ravishankar, 2006; Kugonza, 2009; and Tingek *et al.*, 2009.

General Analysis

In addition to the 20 races of *Apis mellifera* described in Table 1, Dietz (1992) put *A. m. cypria* Pollmann (1879), *A. m. syriaca* Buttel-Reepen (1906), *A. m. meda* Sorikov (1929) and *A. m. armeniaca* Sorikov (1929) in the Near East group. Dietz (1992) therefore produced a list of a total of 24 races of *A. mellifera*. Hussein (2000) added another 2 races; *Apis mellifera abyssinica* and *A. m. sudanesis* in the African group in the countries of Ethiopia and Sudan respectively. Wikipedia (2012) mentioned 2 other races of *A. m. lihzeni* and *A. m. ruttneri* in Western Mediterranean and Northwestern European Group, and Central Mediterranean and Southeastern European Group

A Review of Honey Bee Biodiversity, Behaviour and Management

respectively. Because information about the behavioral and morphological description and geographical distribution of these 8 races is scanty, they have not been put in Table 1. However, this work considers all the 28 races of *Apis mellifera* mentioned so far. *Apis major*, mentioned in Wikipedia (2012), was dropped earlier by Dietz (1992), and consequently not considered here. Although Dietz (1992) put *A. m. carnica* in the Central Mediterranean and Southeastern European Group, Hussein (2000) reported its presence in Africa in Egypt, Libya, Tunisia, Somalia and Sudan. *A. m. ligustica* belonging to the same group was also reported in Africa in Egypt and Libya. Hussein (2000) also reported the presence of *A. m. adansonii* in Kenya, Uganda and Tanzania. The Oriental honey bees in Table II have duly been recognized, although a lot of information is lacking on the description and geographical location of *Apis koschevnikovi*, *A. nigrocinta* and *A. nuluensis*, all descendants of *A. cerana*. Rinderer (1999) reported that *A. florea* has gotten to Africa and Hussein (2000) reported its presence inland Africa in Sudan. The western honey bees, *Apis mellifera*, differentiated into the currently 28 recognized geographic races as they spread from Africa into Eurasia. All the races are cross fertile, but the geographical isolation led to numerous local adaptations such as brood cycles synchronized with the bloom period of local flora, forming a winter cluster in colder climates, migratory swarming and absconding in Africa, enhanced foraging behavior in deserts, overly aggressive/defensive behavior among African honey bee races, and modifications in body structures that may prevent reproduction among the races. These behavioral and structural adaptations dictate upon the method used by man to manage bees, as stated by B/D (2003a) and Nombre, *et al.* (2009). It is worth appreciating that behaviors such as the migratory swarming and absconding, and overly aggressive/defensive and stinging behavior in African honey bees are responses developed over time, partly, to equivalently fierce diseases, parasites, pests and predators. In addition to the diseases and parasites that affect honey bees in general, listed in Table 3, Shimanuki, *et al.* (1992) reported that Bee Virus Y like the Black Queen Cell Virus (BQCV) is associated with *Nosema* and Bee Virus X is associated with amoeba-infected bees. The pathological importance of Cloudy Wing Virus (CWV), Deformed Wing Virus (DWV), Slow Bee Paralysis Virus, Egypt Bee Virus, Kashmir Bee Virus (KBV) and Arkansas Bee Virus (ABV) including the S-shaped Virus reported by Sanford (2009) is unknown. *Spiroplasma* sp., gregarines such as *Monoica apis*, *Apigregarina stammeri*, *Acuta rousseaui* and *Leidyana apis* and flagellates (*Crithidia* spp.) are known to be associated with honey bees, but there is no evidence of them being harmful but can be opportunistic in the presence of a disease such as *Nosema* (Shimanuki, *et al.*, 1992). Chilled brood disease is associated with shortage of adult bees to cover all the brood, during spring or cold weather, whereby chilled brood is found mostly on the fringes of the brood area, with healthy brood at the centre. Chilled larvae or pupae appear yellowish, tinged with black on segmental margins; or brownish or black, crumbly, pasty, or watery. The brood cells are punctured or uncapped, pupae decapitated. It is important to note that decapitation may also result from the feeding of the lesser wax moth larvae. Other diseases include overheated brood disease, starved brood disease, overheated bees, and lethal genes disease. Sanford (2009) described a deleterious purple brood disease in southern US (Madison, Taylor and Jefferson counties of Florida), associated with areas abounding

with summer titi plant (*Cyrilla racemiflora* L.). It is believed that either the nectar and/or pollen of the plant is responsible for killing the brood and turning it purple. Larvae, pupae and newly emerged bees may be affected. Kugonza (2009) also described a condition of an abrupt disappearance of worker bees of *Apis mellifera* from a hive, called Colony Collapse Disorder (CCD), reported in North America, Europe and Taiwan. The cause of this condition is not precisely known. In addition to the bee pests and predators summarized in Table IV, pesticides and insecticides are known to poison bees, as stated earlier in page 5, section 2(iv), and have also been considered here as pests. Geckos, cockroaches and earwigs sitting under the lids of hives are usually no threat and can even help to control intruders (Shimanuki, *et al.*, 1992). However, cockroaches and earwigs can be very offensive when the extractor puts them in the extracted honey! Mould, a fungus, can be seen growing on the inside of the hive and this is caused by excess moisture. Mould may indicate that the hive is sited in an inappropriate damp place. This is a problem mainly during and shortly after the rains but is generally not a major problem. Remove any unattended empty combs from the edges of the nest if bees fail to fan any surplus moisture out of the hive. Increase the number of entrance holes to improve hive ventilation and air passage.

Hoofed animals like the cattle and horses have been reported to bother bees, but can be kept off the hives by fencing the apiary. Baboons and chimpanzees have also been known to hunt honey from bee colonies. Other reported honey bee pests include toads (*Bufo marinus*), cleptoparasitic drone flies and other robber bees. Some mammalian carnivores are also predators of bees including the jackals, raccoons, coatis, martens, some weasels, skunks and the opossum; their control is by trapping the animals or fencing the apiary, and when laws permit, these animals can be trapped and shot. Bears smash bee equipment to obtain brood and honey, and it returns repeatedly to an apiary, often resulting in destruction of the entire apiary in more isolated areas before their damage is detected, which is usually enormous! Bear control can involve removal, aversive conditioning or permitted destruction of bears; use of sturdy electrified fences with 3 to 4 strands, which must continuously be charged; and removal of colonies from the affected areas.

Actualization

Man is striving to conserve and utilize the honey bee resource for its honey, pollen, beeswax, propolis, royal jelly and pollination activities which are all economically important. Beekeeping is therefore practiced in different parts of the world at varying levels. The behavior of the honey bee dictates upon the method used for its management. For example, the cavity-nesting species such as the *Apis mellifera* with all its races, *A. cerana, A. koschevnikovi, A. nigrocinta* and *A. nuluensis* are being kept in hives. They are also relatively mild. The European honey bee species, *Apis mellifera*, has been extensively researched to the extent that the genome is known and it is now a highly developed and domesticated bee. The Asian honey bee species, *Apis dorsata, A. laboriosa, A. florae* and *A. andreniformis*, are open-nesting and are still being exploited from mountain cliffs and tree branches in the wild, although attempts have been made recently to hive *A. florea* in Oman (Whitcombe, 1985). The African *Apis mellifera* honey

bee and its races are also cavity-nesting and being kept in hives, but are overly aggressive/defensive with high propensity to swarm, migrate and abscond. Although all the above mentioned honey bee species sting and people fear bee stings world over, the African honey bee, particularly *Apis mellifera scutellata*, is the most defensive bee race and can attack a human or an animal with 500 to 5000 stings! This behavior makes its management more difficult. Fortunately there are related but gentler races like the *Apis mellifera monticola* available in the neighboring ecosystems in the East African Mountains. Research is now focusing on how to tap these genes of gentility to improve honey bees of Africa. Across the board, there are management challenges posed by honey bee diseases and parasites, and Table 3 below summarizes honey bee parasites and diseases together with their geographical distribution, symptoms and management measures available. Table IV summarizes the various honey bee pests and predators, their distribution, damages they cause and how they are managed. In Table 5 pesticides which are highly toxic to bees are pointed out. These pesticides are not to be used within or around apiaries, as they will kill the bees. Moderately toxic and relatively nontoxic pesticides are listed in Tables VI and 7 respectively.

Table: 3: A summary of Bee Diseases and Parasites, their Geographical Distribution, Symptoms and Management Measures:

S/NO.	Disease	Causative agent and its geographical distribution	Symptoms	Management measures
BROOD DISEASES				
A. Viral Diseases				
1.	Sacbrood	Sacbrood virus (SBV), *Morator aetatalus*: Reported present in Europe, China, India, USSR, Papua New Guinea, Australia, New Zealand, Brazil, Mexico, US, Canada, Algeria, Egypt, Tunisia and South Africa.	Attacks older larvae before and after capping; few abnormal cells but almost no torn cappings appear on the brood comb; partially uncapped cells scattered among the capped brood or capped cells that remain after surrounding brood has emerged; affected larvae change from pearly white to gray and finally black; affected larvae remain in partially or capped cells; the head region darker than the rest of the body; when carefully removed from their cells, the affected larvae appear like whitish water-filled sacs, hence the name; no characteristic odor; typically, the scales are brittle but easy to remove	Concentration of infected colonies, including burning of infected brood combs; chemotherapy by dusting brood nest with oxytetracycline powdered sugar; thorough disinfection of contaminated equipment; and queen exchange may help.
2.	Black queen cell (BQC)	black queen cell virus (BQCV): Reported in Uganda and parts of the world	Mostly affects housed apiaries and those in damp places; the infected queen larva turns black and dies; associated with *Nosema apis*.	Avoid siting hives in damp places; select sites with good air circulation for increased ventilation; remove the affected hive

A Review of Honey Bee Biodiversity, Behaviour and Management

				from the apiary; remove the dead bee; treat for *Nosema*
B. Fungal Diseases				
3.	Chalkbrood	*Ascosphaera apis:* Reported present in Argentina, Mexico, US (Florida), Canada, Northwestern Europe, China, Algeria, Egypt, Libya and Tunisia	Mildly pathogenic spores which can survive over 15 years; infests the gut of the larva; competes with the larva for food, causing it to starve; continues to consume the rest of the larva's body, causing it to appear white and chalky, hence the name; associated with cold temperatures and colony stress, hence a "stress disease"	Maintain a strong healthy colony; increase ventilation through the hive and/or re-queen the hive.
4.	Stonebrood	*Aspergillus flavus* or *A. fumigatus and A. niger:* World-wide distribution, including Algeria, Egypt, Libya and Tunisia	The fungi are soil inhabitants; there is a characteristic whitish-yellow collar-like ring near the head end of the infected larva; after death, the affected larva becomes hardened and difficult to crush, hence the name stonebrood; fungus erupts from the integument of the larva and forms a false skin (mummy); also pathogenic to adult bees and other insects, as well as mammals and birds.	Maintain strong colonies
C. Bacterial Diseases				
5.	American foulbrood (AFB)	*Bacillus larvae larvae:* worldwide in distribution; Uruguay, Mexico, US, Canada, Algeria, Libya, Morocco, Tunisia, Europe, USSR, Iran, China, India, Japan, Australia and New Zealand	Most widespread and destructive of all bee brood diseases; most difficult to eliminate completely; relatively high morbidity; young larvae less than 24 hours old most susceptible; infected larvae darken and die after their cell is sealed; dark, viscous and ropy pus, with decayed gluepot odor; the individual later becomes a black and hard scale, glued tight to the cell; millions of spores are produced.	Brood nest inspection at least 3 times in a year for normal population-brood food stores balance; introduction of a minimum of 8-10 foundations into brood nest every year; removal of infected brood combs and concentration of strong and infected colonies; treat with sulfathiazole and oxytetracycline hydrochloride (Terramycin); instant destruction (burning) of all diseased weak and failing colonies;

				and thorough disinfection of contaminated equipment by boiling in 2% NaOH for 15 minutes.
6.	European foulbrood (EFB)	*Melissococus pluton:* worldwide in distribution; Brazil, Paraguay, Peru, Columbia, Chile, Argentina, US, South Africa, Angola, Zambia, Malawi, Tanzania, Madagascar, Algeria, Libya, Tunisia, Morocco, Saudi Arabia, Iran, Europe, India, China and Australia.	Mildly pathogenic spores which can survive 2-3 years only; infected larvae die before they are capped; leathery larval pieces in watery pus; sour smell; larvae up to 5 days old can be seen to twist in un-natural positions; brood becomes spotted; the individual later becomes brown, soft and loose in the cell; and bee population declines	Concentration of weak colonies, including burning of infected brood combs; dusting the brood nest with oxytetracycline powdered sugar; and thorough disinfection of contaminated equipment.
ADULT BEE DISEASES				
A. Viral diseases				
7.	Acute bee paralysis (ABP)	Acute bee paralysis virus (ABPV or APV): Belongs to the family Dicistroviridae as Israel acute paralysis virus (IAPV), Kashmir bee virus (KBV), and the black queen cell virus (BQCV): Reported in Europe, North America, and Antiqua	Frequently detected in healthy colonies; associated with sudden collapse of colonies infested with *Varroa destructor.*	Treat for *Varroa*
8.	Chronic bee	Chronic bee paralysis virus	Occurs in seemingly healthy bees; infected worker bees become hairless	No control and treatment methods

	paralysis (CBP)	(CBPV): Reported in Britain, Australia, North America, Europe, and USSR	and develop a uniform black color	reported
B. Protozoan Diseases				
9.	Nosemosis	*Nosema apis or N. ceranae*: Worldwide; reported present in Brazil, Uruguay, Paraguay, Argentina, Chile, Bolivia, Mexico, US, Canada, South Africa, Zimbabwe, Tanzania, Algeria, Egypt, Libya, Tunisia, Morocco, Senegal, Kenya, Ethiopia, Europe, USSR, Iran, India, China, Papua New Guinea, Australia and New Zealand	Develops when bees can not leave the hive to eliminate waste for a long period especially during cold spell or when confined for a purpose, or when there is rapid brood building or nutritional imbalance; there is dysentery; disjointed wings; distended abdomens; absence of stinging reflex; stupefied, disoriented or paralyzed behavior (bees become weak and may crawl around the front of the hive, unable to fly); reduction in honey crop; accelerated queen supercedure; also associated with BQCV.	Selecting hive sites with good air flow to increase ventilation through the hive; avoid damp and cold places; treat with antibiotics e.g. by feeding the bees with Fumidil B® in sugar syrup; fumigate infected equipment with acetic acid; and sterilize contaminated hive equipment with elevated temperatures of up to 49^{0}C for 24 hours.
10.	Amoeba disease	*Malpighamoeba mellificae*: Reported in Europe, the Soviet Union, North America (Mexico, US and Alaska), South America (Argentina and Venezuela), New Zealand, China, Australia, Algeria, Tunisia	"Spring dwindling" or "disappearing" condition; often mixed with *N. apis*; shiny swollen malpighian tubules; queens may be infected	Decontamination of equipment with acetic acid

A Review of Honey Bee Biodiversity, Behaviour and Management

		and Ethiopia		
C. Parasitic mites				
11.	Acariosis	*Acarapis woodi* (other similar species: *A. externus and A. dorsalis*): Reported in Europe, USSR, Iran, India, China, Japan, Madagascar, Mozambique, Egypt, Algeria, Tunisia, Morocco, Brazil, Uruguay, Paraguay, Chile, Columbia and Mexico.	Infected bees become weak, crawl at the hive entrance and sometimes uncouple their wings so that all the 4 wings are visible; death occurs from disruption to respiration due to the mites clogging the tracheae, damage to the tracheae and secondary infections; paralysis of the colony; reduction in honey production; likelihood of winter survival decreases with increasing infestation	Treat infested colonies with Miticur® or special formulations of menthol
12.	Varrosis	*Varroa destructor* and *Varroa jacobsoni*: Worldwide; reported in Europe; Southeast Asia including India, Hong Kong and Philippines, New Zealand; US including Florida, Texas, Mexico; Brazil; Africa including Algeria, Egypt, Morocco and Tunisia	Attack and suck blood of older larvae, pupae and adults (drone brood preferred to worker brood); decrease in vitality of emerging bees; shortened abdomen; deformed legs; death of pupae; infested colonies die off within 3-4 years unless treated; *Varroa* also vector virus which further compounds the situation; bees infected with this virus during their development will often have visibly deformed wings.	Treat with Apistan®, a formulation of fluvalinate or formic acid gel packs; and development and use of honey bee stocks resistant to V*arroa*
13.		*Tropilaelaps clareae, T. koenigerum,* and *Euvarroa sinhai*, external mites: Reported in Southeast Asia including	Attack both developing brood and adult honey bees; associated with *Varroa jacobsoni* infection	Treat like for *Varroa* with fluvalinate and amitrax; and break the brood cycle.

		India, Iran, Pakistan, Afghanistan and Papua New Guinea			

Sources: Kshirsagar and Phadke, 1985; Rashad, *et al.*, 1985; Akratanakul, 1985; Xianshu, 1985; Koeniger, N and Koeniger, G., 1985; Woyke, 1985; Ingemar, 1999; B*f*D, 2003c; Shah, 2006; Hussein, 2000; Kugonza, 2009; B*f*D, 2010; and MAAIF, 2012a.

Table: IV: A summary of Bee Pests and Predators; the Damage they cause and their Geographical Distribution and Management Measures:

S/NO.	Pest/Predator	Geographical Distribution	Damage caused	Management Measures
	Pests			
1.	Spiders and dragon flies	World-wide, including Uganda	Capture foraging honey bees. However, the damage to the colony is insignificant	No control measure reported
2.	Strepsipteran insect, *Stylops melittae*	Not widespread	The female positions itself between the abdominal segments of the bee and feeds on the blood of the bee. Damage, insignificant.	No control measure reported
3.	Hive beetles: Small hive beetle (SHB), *Aethina tumida*; and the Large Hive Beetle (LHB), *Oplostomus fulgineus*	Africa including Botswana, Tanzania, Nigeria, Senegal and Uganda; and US	The SHB lays eggs in pollen cells, which can be turned into a stinging mess by the maggots within few days; both the adult and larvae feed on stored pollen and honey, damage combs; and brood is killed by burrowing beetles; honey ferments and bubbles out of the cells smelling like decaying oranges; bees abandon heavily infested hives. Damage caused is only comparable with that of wax moths. The LHB feeds on brood and is most numerous during the rains.	Hand pick or destroy the beetles. Use holes instead of slits as hive entrance and reduce the size of the entrance holes. Use of wood preservatives or a barrier of grease or oil on the stands and the suspension wires of the hive. Regular inspection and good management and hygiene practices. Maintain strong colonies. Keep hives well ventilated and shaded
4.	Bee louse (*Braula coeca*)	World-wide: Reported in: Africa including Congo, Egypt,	Steals food from the mouth of the honey bee, *Apis mellifera*, usually the royal jelly	Remove the cell cappings and extract and sell liquid honey

A Review of Honey Bee Biodiversity, Behaviour and Management

		Morocco, Algeria, Tunisia, Kenya, Tanzania, Uganda, Ethiopia, Somalia, Guinea Bissau, Nigeria and Senegal; India and China in Asia; the Soviet Union; Tasmania in Australia; Europe; and the Americas	from the queen bee; the larvae burrow under the cell cappings, thereby destroying the appearance of the combed honey to be sold directly in the market.	
5.	The Greater Wax Moth (GWM), *Galleria mellonella* (L.)	Warm areas of the world, including Uganda, Algeria, Egypt, Libya, Kenya, Tanzania, Sudan, Senegal, Southeastern and Southwestern US including Florida, Asia including Jamaica	Adult moths lay eggs near wax combs; the larvae hatch and feed on unprotected combs, cast larval skins, pollen and honey; bore into and leave silk-lined tunnels or galleries in the comb, sometimes reducing it to just a mass of web, thereby damaging it considerably	Use of Certan® containing spores of selected strains of *Bacillus thuringiensis*, as a biological control agent; treatment with Paradichlorobenzene (PDB), but the eggs are not affected; and cold treatment and fumigation with Carbon dioxide; maintain a strong colony.
6.	The Lesser Wax Moth (LWM), *Achroia grisella* and others like *Aphomia sociella*	Algeria, Egypt, Kenya, Tanzania, Uganda, Ethiopia, Sudan, Ghana, Nigeria and Senegal	The LWM tends to attack processed wax.	Always use clean wax on starter strips! Scrape away any eggs, wax moth faeces or pupa. Kill any larvae or adult wax moths. Plug any holes and cracks in the top bars and the hive body. Put PDB crystals on top of the top bars and cover with a lid. Remove old combs during times of food scarcity when the colony size shrinks. Keep the hive clean and free from bits of comb and debris. Maintain a strong colony. Unite weak colonies with stronger ones.
7.	Termites	World-wide, including Jamaica, Uganda, Kenya, Tanzania	Termites attack wooden hive types and equipment; damage can be severe in subtropical and tropical world	Suspend hives between trees instead of poles. Alternatively poles with used engine oil and place the supports in tins of oil. Also avoid using unburnt bricks as hive supports.
8.	Squirrels, mice, dormice, rats and lizards	Somalia, Tanzania, Uganda, Sudan, Ghana, Nigeria	Build nest in the hive; destroy honey combs; and the smell of their	Reduce the size of hive entrances; keep the animals out of stored equipment; protect

A Review of Honey Bee Biodiversity, Behaviour and Management

		and Senegal	droppings will discourage swarms from occupying the hive; and may even damage the hive altogether.	unoccupied hives against pests and keep them clean and baited; and protect the hive by enclosing it in a wire screen of mesh of size small enough to allow only the bees to pass through.
	Predators			
9.	Ants: Red ants (*Camponotus abdominalis*); black ants (*Cremastogaster brevispinosa*); and brown house ants (*Pheidole megacephala*); and others including carpenter ants, both large and small	World-wide, including Uganda, Algeria, Egypt, Morocco, Kenya, Tanzania, Ethiopia, Sudan, Nigeria, Senegal, Jamaica, Botswana, US including Florida	Eat nectar, honey, brood and the bee's body; and live in hollows like the bee itself and the hive. May overrun a colony	Use of wood preservatives or a barrier of grease or oil on the stands and the suspension wires of the hive; spreading wood ash or charcoal ash or Johnson baby powder around the stands of the hive or tying on rags dipped in diesel or petrol will also keep ants away; maintain strong colonies; regular inspection and good management and hygiene practices; seek their nests and deal with them at the source
10.	Honey badger or horr (*Mellivora capensis*)	Somalia, Kenya, Botswana, Tanzania, Uganda and Ethiopia,	Breaks into hives to eat honey and brood. This can be total damage.	Hang hives securely one and a half metres from the ground; or, suspend the hives; tie lids on the hives securely with wire and put heavy stones on the covers
11.	Man	World-wide including Kenya, Uganda, Tanzania and Oman	There is a lot of damage caused to hive colonies by man through theft, vandalism and setting bushfires that burn them. There is also reduction of bee forage by large scale agriculture and urbanization. Man is probably the most dangerous species and the worst enemy to bees	Keep hives where they can be supervised. Place hives in a house or within a strong fence placed around the apiary, with a lockable gate. Encourage and enforce good neighbor policy
12.	Birds: Honey guide (*Indicator indicator*) and bee eater (*Merops species*)	Algeria, Egypt, Libya, Kenya, Uganda, Tanzania, Ethiopia, Nigeria, Senegal and Jamaica	The honey guide leads mammals like honey badgers, baboons and man to the honey bee nest. The larger animal digs out and exposes	Don't leave any brood combs exposed and scare the birds away. Alternatively place your hives in a bee house. In case of the honey guide bird, you may have to move the honey bee

A Review of Honey Bee Biodiversity, Behaviour and Management

19

			the bee nest and the bird rewards itself by eating mainly the young bees, brood and the entire combs. It can also capture bees flying out of the hive. The merops eat honey bees and other venomous Hymenoptera almost exclusively.	colonies away from the area or control the honey badger
13.	Wasps (e.g. *Vespula vulgaris*), including the yellowjackets, the banded bee pirate, *Palarus latifrons*, and the bee pirate, *Philanthus triangulum*	US including Florida; Algeria; Egypt; Libya; Senegal; Kenya; Tanzania; Uganda; Sudan; Botswana and South Africa	Molest colonies, attack and eat worker bees. May overrun a colony	Fill any gaps and holes in the hive and make entrances small enough for the bees to defend themselves. Cut a bottle with a narrow mouth in half and invert the top, put some water and jam in; the wasps will enter and drown. Do not bait with honey, as bees will also be killed. A meat trap can also be used. Seek their nests and deal with them at the source. *Palarus latifrons* can be controlled using a 'South African trap', made by immersing a mirror in a basin of water which has a thin layer of paraffin on top, and put in front of the hive when bees are not foraging. Fly traps can also be used.

Sources: Clauss, 1985; Corner, 1985; Phokedi, 1985; Shimanuki, *et al.*, 1992; Ingemar, 1999; Paterson, 1999; Raina, *et al.*, 1999; Hussein, 2000; B*f*D, 2003a; Sanford, 2009; Kugonza, 2009; MAAIF, 2012a; MAAIF, 2012b; and, Allan and Hardie, http://www.honeybee.com.au/

Table 5: A summary of some pesticides highly toxic to bees (LD$_{50}$: 0.001-1.99 µg/bee):

S/NO.	Pesticide group	Poisoning symptoms	Pesticide examples	LD$_{50}$	Slope value, Probits
1.	Organophosphorous pesticides	Regurgitation (bees are wet), bees are often disoriented, lackadaisical and remain in the hive awaiting paralysis and death, erratic self cleaning behavior, tumbling about, wings held away from the body but usually remaining hooked together, distended abdomen, queens cease egg laying	Parathion	0.175	4.96
			Methyl parathion EC	0.111	5.13
			Methyl parathion, Penncap-M®	0.348	5.38
			Fenitrothion	0.176	5.75
			Dimethoate	0.191	5.84
			Methamidophos		
			Methidathion	0.237	8.48
			Mevinphos	0.305	7.77
			Tepp	0.002	0.68

A Review of Honey Bee Biodiversity, Behaviour and Management

S/NO.	Pesticide group	Poisoning symptoms	Pesticide examples	LD₅₀	Slope value, Probits
			Diazinon	0.372	8.03
			Dichlorvos	0.501	8.61
			Malathion	0.726	7.83
			Chlorpyrifos (Dursban and Lorsban)	0.110	10.17
2.	Carbamate pesticides	Aggressiveness, erratic movements, unable to fly, stupefied as though they had been chilled, paralysis, moribundity and death, queens often cease egg laying and hive bees often initiate Supersedure queen cells before egg laying resumes, most poisoned bees die at the colony	Carbofuran, Furadan®	0.149	6.14
			Mexacarbate, Zectran®	0.302	4.87
			Carbosulfan, Advantage®	0.362	4.70
			Carbaryl	1.54	3.04
3.	Pyrethroid, pyrethrum and botanical pesticides	Direct contact exposure may cause regurgitation, erratic movements, unable to fly, stupefaction, paralysis, moribundity and death within a very short period of time, some bees die in the foraging area, others die between the foraging area and the hive, and, the remainder die at the hive.	Bifethrin	0.016	3.06
			Cyfluthrin	0.029	3.49
			Cyhalothrin	0.052	3.32
			Cypermethrin	0.060	1.95
			Decamethrin, Decis®	0.067	4.88
			Flucythrinate (Pay-off)	0.078	3.73
			Fenpropathrin (Danitol)	0.120	2.97
4.	Chlorinated Hydrocarbon pesticides	Erratic movements, abnormal activities, trembling, dragging the hind legs as if paralyzed, wings held away from the body but usually remaining hooked together, many bees able to fly to the field until shortly before morbidity occurs, a high percentage of poisoned bees die in the field and between the field and colony as well as at the colony	Dieldrin	0.133	2.51
			Aldrin	0.352	5.06
			Heptachlor	0.526	5.94
			Lindane	0.562	5.07

Source: Atkins, 1992

Table 6: A summary of some pesticides moderately toxic to bees (LD₅₀: 2.0-10.99 µg/bee):

S/NO.	Pesticide group	Poisoning symptoms	Pesticide examples	LD₅₀	Slope value, Probits
	Chlorinated Hydrocarbon pesticides	Erratic movements, abnormal activities, trembling, dragging the hind legs as if paralyzed, wings held away from the body but usually remaining hooked together, many bees able to fly to the field until shortly before morbidity occurs, a high percentage of poisoned bees die in the field and between the field and colony as well as at the colony	Endrin	2.02	4.20
			DDT	5.95	4.89

Source: Atkins, 1992

A Review of Honey Bee Biodiversity, Behaviour and Management

Table 7: A summary of some pesticides relatively nontoxic to bees:

Pesticide	Pesticide
Acaricides, Diseases, Insect Growth Regulators (IGRs) and Insecticides	
1. Allethrin	18. Esfenvalerate
2. Aldoxycarb, Standak®	19. Ethion
3. Amitraz	20. Methoxychlor
4. Azadirachtin	21. Multimethylalkenols
5. *Bacillus thuringiensis*, Biotrol®, Dipel®, Thuricide®	22. Nicotine
6. *B. t., Kurstaki*, Javelin®, Dipel®	23. *Nosema locustae* Canning
7. *B. t., tenebrionis*	24. Oxythioquinox
8. Chlordimeform	25. Pirimicarb
9. Chlorobenzilate	26. Polynactins
10. Cryolite	27. Propargite
11. Clofentizine	28. Pyrethrum
12. Cymiazole	29. Pyriproxyfen
13. Cyromazine	30. Rotenone
14. Dibromochloropropane	31. Tetradifon
15. Dicofol	32. Tetraflubenzuron
16. Diflubezuron	33. Trichorfon
17. Dinobuton	Z-11-hexadecanol, tomato pinworm pheromone
Fungicides	
1. Anilazine	15. Fenaminosulf
2. Benomyl	16. Folpet
3. Bordeaux mixture	17. Glyodin
4. Captafol	18. Maneb
5. Captan	19. Nabam
6. Copper oxychloride sulfate	20. Polyphase™ P-100
7. Copper 8-quinolinate	21. Prochloraz
8. Copper sulfate (monohydrated)	22. Prochloraz/carbendazin
9. Cuprous oxide	23. Sulfur
10. Dazomet	24. Thiram
11. Diniconazole	25. Triforine
12. Dinocap	26. Triphenyltin hydroxide
13. Dithianon	27. Ziram
14. Dodine	
Herbicides, Defoliants, Dessicants and Plant Growth Regulators (PGRs)	
1. Alachlor	27. Methazole
2. Amitrole	28. Metribuzin
3. Atrazine	29. Monuron
4. Bentazone	30. Naptalam
5. Bromacil	31. Nitrofen
6. Butifos	32. Norflurazon
7. Chlorbromuron	33. Paraquat
8. Chloroxuron	34. Phenmedipham
9. Cyanazine	35. Picloram
10. Dalapon	36. Prometryn
11. DEF®	37. Pronamide

A Review of Honey Bee Biodiversity, Behaviour and Management

12. Dicamba	38. Propanil
13. Dichlobenil	39. Propazine
14. Diquat	40. Propham
15. Diuron	41. Quinchlorac
16. EPTC, Eptam[®]	42. Simizine
17. Ethalfluralin	43. Sodium chlorate
18. Etephon	44. Terbacil
19. EXD, Herbisan[®]	45. Terbutryn
20. Fluometuron	46. Thiadiazuron
21. Fluridone	47. Tribuphos
22. Hydrogen cyanamide	48. Uniconazole
23. Imadagylin, Arsenal[®]	49. 2, 3, 6-TBA
24. Linuron	50. 2, 4-D
25. MCPA, Mapica[®]	51. 2, 4-DB
26. Metaldehyde	52. 2, 4, 5-T

Source: Atkins, 1992

Discussions

Globally, a lot of effort has been put to address constraints facing beekeeping industry with varying degrees of success in different parts of the world. The success of the efforts to sustainably harness and exploit the honey bee resource depends largely and firstly on the precise identity and knowledge of the species and the race in question, its characteristic behavior and the factors affecting the behavior. Then, it is this knowledge in totality that should be applied to develop appropriate technologies to manage the honey bee resource profitably in a sustainable manner. Unfortunately, there is apparent confusion on the existence and exact distribution patterns for particularly the *A. m. adansonii* and *A. m. scutellata* honey bees in East Africa. Originally, all honey bees south of the Sahara were called *A. m. adansonii* but Dietz (1992) removed *A.m. adansonii* from the rest of the south Sahara and restricted the name to the bees of the African west coast. Although authors like Ruttner and Kauhausen (1985), Mbaya (1985), Mbae (1999), Coleman (1999), Raina, *et al.* (1999), Rinderer (1999) and Kugonza (2009) agree with Dietz (1992) stating that *A. m. scutellata* covers the central and eastern part of African continent, from Ethiopia to the Great Karoo in South Africa, including Uganda, Kenya and Tanzania, Hussein (2000) has continued to restore *A. m. adansonii* to Uganda, Kenya and Tanzania, and removing *A. m. scutellata* completely from Uganda and Tanzania. MAAIF (2012a) was silent about the honey bee species in Uganda and called for research to find it out. Whether Hussein (2000) was right or not, most of the information on East African honey bees, particularly of Uganda, has been based on limited sampling or generalizations of researches done mainly along the coast of Kenya and Tanzania. So information on exhaustive surveys and researches done and widely published specifically on Ugandan honey bees is scanty. Raina, *et al.* (1999) had made a similar observation but on the lack of documentation, particularly, on honey bee pests and diseases in Africa as a whole, blaming the infancy of beekeeping industry on the continent and lack of research. The hope now is the recent establishment of Apiculture Research Unit under NARO at NaLIRRI, Tororo in Uganda. On the other hand, African honey bee industry was evidently rushed into exogenous technology, without heeding to the indigenous

knowledge. Concerned indigenous authorities, professional bodies, research institutions and universities appeared to have paid less attention to the very important resource-the honey bee which works hard daily to pollinate flowering cultivated crops and wild plants for the benefit of man and other living creatures! It has just been realized recently that among the so called hostile African honey bees, there are some gentle and productive races. This mixture of varying capabilities among the African bees is a very good genetic preserve for further research and breeding of better stocks on the African continent. It is therefore apparent that the problem of the African honey bee management was blown out of proportions for nothing. It is possible now to select and breed a superior stock of honey bee from within the races that are available on the continent. This calls for commitment from the concerned indigenous authorities, professional bodies, research institutions and universities, of course with collaboration with relevant international bodies.

General Recommendations

Exhaustive field surveys and research on indigenous bee species and races, including stingless bees, should continuously be carried out in Uganda and other African countries to document and update national bee information database. This is very important in the face of exotic species and diseases versus Biosecurity; for early detection and warning.

Deliberate bee breeding programmes should be initiated to develop gentle and productive honey bees from within the African bee species and races.

African Governments should actively fund and develop beekeeping industry in their countries as it is one of the agricultural enterprises a rural poor household can afford and actively engage in to earn their livelihoods.

Already established commercially important bee forage plants should be developed and cultivated specifically for bees (api-forestry) to help reduce incidences of swarming, migration or nomadism and absconding in all African countries; and this is an area that has also been neglected for a long time in beekeeping industry in many countries in the world.

There should be a systematic and continuous search for indigenous bee forage plants other than those already established commercially important ones, most of which may be exotic. The indigenous plants, if developed, may establish themselves better in the local environment and be more resilient to climate change and the resultant effects like drought, floods and storms.

Kajobe (2007) established presence of at least 5 stingless bee species in Bwindi Impenetrable National Park in Uganda; including *Meliponula bocandei*, *M. nebulata*, *M. ferruginea*, and *Hypotrigona gribodei*. Keeping of stingless bees (Meliponiculture) should therefore be introduced and promoted in Uganda as an alternative to the honey bees.

Conclusion

The species of honey bees existing in some African countries like Uganda, to date, are not precisely known. This is likely to affect development of appropriate technologies for honey bee management as the process will be misguided. It is also likely to affect problem solving and decision making in case of accidental introduction or natural spread into the country of a foreign honey bee, honey bee pest or disease.

The greatest constraint facing the management of the African honey bee is its overly aggressive/defensive and stinging behavior and high rates of swarming and absconding which many people including farmers can not cope with. It has and will continue to frustrate well intentioned beekeeping projects in Africa.

Beekeeping is a forest-based enterprise and gives both ecological conservation and socio economic benefits to the community. It can easily be taken up by the rural communities along forests.

Literature available points to the possible existence of *Apis mellifera scutellata* and *Apis mellifera adansonii* in Uganda. It is not yet confirmed whether Uganda has either *Apis mellifera scutellata* or *Apis mellifera adansonii* or both. No other particular honey bee species has been established in Uganda. It is likely that Uganda could be having both races, plus others. This state of confusion reflects lack of deliberate field surveys and research on the indigenous honey bees in Uganda. It may also be reflecting lack of wide publications of such research findings, if any. This, clearly, has highlighted the enormous opportunity for further field based research on the indigenous honey bee resources and to confirm which honey bee species exist in Uganda.

Bibliography

- AHMAD, F.; JOSHI, S.; and GURUNG, M. "Information from ICIMOD: Conserving wild bees and promoting *Apis cerana* beekeeping". B/D Journal. Issue No. 67 (June 2003): p10.
- AKRATANAKUL, P. "Apiculture and Bee Management Problems in Tropical Asia, America and the Pacific". Proceedings of the Third International Conference on Apiculture in Tropical Climates. Nairobi, Kenya. 5-9 November 1984. IBRA. 1985: p89-92.
- ALLAN, L. and HARDIE, D. Bee Pests and Diseases. http://www.honeybee.com.au/
- ATKINS, E. L. "Injury to honey bees by poisoning". The Hive and the Honey Bee. Dadant & Sons, Inc. Hamilton, Illinois. 1992: p1153-1208.
- B/D (2003a). "Zoom in on Cote D'Ivoire". B/D Journal. Issue No. 66 (March 2003): p6.
- B/D (2003b). "News around the world". B/D Journal. Issue No. 67 (June 2003): p12-13.
- B/D (2003c). Bees and Rural Livelihoods. Bees *for* Development. Troy. Monmouth, UK. 2003.

- B/D (2010). Organic Beekeeping- A Discussion. BfD Journal. Issue No. 96 (September 2010): p6-7.
- BRAGA, J. A.; NUNES, R. M.; and LORENZON, M. C. A. "Stingless Bees as Bioindicators in Brazil". B/D Journal. Issue No. 92 (September 2009): p7-9.
- CLARE, T. (2007). Is there a best bee? A lecture at the 76th National Honey Show 2007, 18th-20th October 2007, Royal Air Force (RAF) museum, London.
- CLAUSS, B. "The status of the banded bee pirate, *Palarus latifrons*, as a honey bee predator in Southern Africa". Proceedings of the Third International Conference on Apiculture in Tropical Climates. Nairobi, Kenya. 5-9 November 1984. IBRA. 1985: p157-159.
- COLEMAN, C. J. "Pollination services by African bees". The Conservation and Utilization of Commercial Insects. Proceedings of the First International Workshop on the conservation and utilization of commercial insects held in Nairobi, Kenya, 18-21, August 1997. Eds Raina, S.K.; Kioko, E.N.; and Mwanycky, S.W. ICIPE Science Press. 1999: p107-114.
- CORNER, J. "Apiculture and bee management problems in African countries". Proceedings of the Third International Conference on Apiculture in Tropical Climates. Nairobi, Kenya. 5-9 November 1984. IBRA. 1985: p41-44.
- CRANE, E. "Tropical apiculture and the need for a global strategy". Proceedings of the Third International Conference on Apiculture in Tropical Climates. Nairobi, Kenya. 5-9 November 1984. IBRA. 1985: p27-32.
- DIETZ, A. "Honey bees of the world". The Hive and the Honey Bee. Dadant & Sons, Inc. Hamilton, Illinois. 1992: p23-61.
- FAJARDO, A. C.; and CERVANCIA, C. R. "Simple ways to manage stingless bees". B/D Journal. Issue No. 67 (June 2003): p3-5.
- FERGUSON, A. W. "The use of pheromones to control honey bee colony behaviour in the tropics". Proceedings of the Third International Conference on Apiculture in Tropical Climates. Nairobi, Kenya. 5-9 November 1984. IBRA. 1985: p123-126.
- GARY, N. E. "Activities and Behavior of Honey Bees". The Hive and the Honey Bee. Dadant & Sons, Inc. Hamilton, Illinois. 1992: p269-372.
- http://en.wikipedia.org/wiki/Apis_mellifera
- HUSSEIN, M. H. "Beekeeping in Africa". Apiacta. 1/2000: p 32 – 48. International Federation of Beekeepers' Associations. http://www.beekeeping.cm/apiacta/beekeeping_africa.htm
- INGEMAR, F. "Techniques for Identification and Prevention of Honey Bee Diseases and Pests". The Conservation and Utilization of Commercial Insects. Proceedings of the First International Workshop on the conservation and utilization of commercial insects held in Nairobi, Kenya, 18-21, August 1997. Eds Raina, S.K.; Kioko, E.N.; and Mwanycky, S.W. ICIPE Science Press. 1999: p75-79.
- KAJOBE, R. "Nesting biology of equatorial Afrotropical stingless bees (Apidae: Meliponini) in Bwindi Impenetrable National Park". Journal of Apicultural Research and Bee World. 46(4). 2007: p245-255.
- KOENIGER, N. and KOENIGER, G. "Change of host by parasitic mites in Asia after a new honey bee species is introduced". Proceedings of the Third International

Conference on Apiculture in Tropical Climates. Nairobi, Kenya. 5-9 November 1984.
IBRA. 1985: p160-162.

- KSHIRSAGAR, K. K. and PHADKE, R. P. "Occurrence and Spread of Thai
  Sacbrood in *Apis cerana*". Proceedings of the Third International Conference on
  Apiculture in Tropical Climates. Nairobi, Kenya. 5-9 November 1984. IBRA. 1985:
  p149-151.
- KUGONZA, D. R. (2009). Beekeeping: Theory and Practice. Makerere University
  and Fountain Publishers Ltd. Kampala, Uganda.
- MAAIF (2012a). The National Beekeeping Training and Extension Manual. Eds
  Kangave, A; Butele, C. A.; Onzoma, A.; and Kato, A. Entebbe, Uganda. 2012.
- MAAIF (2012b). The Strategy to Increase the Rate of Colonization and Number of
  Farmer Preferred Bee Hives, March 2012-June 2017. Entebbe, Uganda. 2012.
- MBAE, R. M. "Overview of Beekeeping Development in Kenya". The Conservation
  and Utilization of Commercial Insects. Proceedings of the First International
  Workshop on the conservation and utilization of commercial insects held in Nairobi,
  Kenya, 18-21, August 1997. Eds Raina, S.K.; Kioko, E.N.; and Mwanycky, S.W.
  ICIPE Science Press. 1999: p103-105.
- MBAYA, J.S.K. "The distribution of African honey bees in Kenya, and some aspects
  of their behavior". Proceedings of the Third International Conference on Apiculture
  in Tropical Climates. Nairobi, Kenya. 5-9 November 1984. IBRA. 1985: p56-60.
- MWANGI, R. W. "Reasons for the low occupancy of hives in Kenya". Proceedings
  of the Third International Conference on Apiculture in Tropical Climates. Nairobi,
  Kenya. 5-9 November 1984. IBRA. 1985: p61-63.
- NOMBRE, I.; BOUSSIM, J. I.; and SCHWEITZER, P. "Improved Top-Bars". B/D
  Journal. Issue No. 92 (September 2009): p5-6.
- OLDROYD, B. P. and WONGSIRI, S. "Asian Honey Bees: Biology, Conservation
  and Human Interactions". B/D Journal. Issue No. 79 (June 2006): p14.
- PALIWAL, G. N.; PALIWAL, S.; and TEMBHARE, D. B. "Practical Beekeeping:
  Eco-friendly harvesting of rock bees". B/D Journal. Issue No. 77 (December 2005):
  p3-4.
- PATERSON, P. D. "Constraints in Transforming Traditional to Modern Beekeeping
  in Kenya". The Conservation and Utilization of Commercial Insects. Proceedings of
  the First International Workshop on the conservation and utilization of commercial
  insects held in Nairobi, Kenya, 18-21, August 1997. Eds Raina, S.K.; Kioko, E.N.;
  and Mwanycky, S.W. ICIPE Science Press. 1999: p95-102.
- PHOKEDI, K. M. "Apiculture and its problems in Botswana". Proceedings of the
  Third International Conference on Apiculture in Tropical Climates. Nairobi, Kenya.
  5-9 November 1984. IBRA. 1985: p64-65.
- RAI, M. M. "Behavioural Ecology of the Wild Honey bee, *Apis dorsata* and its
  Impact on Economic Development in Vidarbha, India". The Conservation and
  Utilization of Commercial Insects. Proceedings of the First International Workshop
  on the conservation and utilization of commercial insects held in Nairobi, Kenya, 18-
  21, August 1997. Eds Raina, S.K.; Kioko, E.N.; and Mwanycky, S.W. ICIPE Science
  Press. 1999: p81.
- RAINA, S. K.; KIOKO, E.; ADOLKAR, V. V.; MUIRU, H.; KIMBU, D.; WEI, S.;
  OUMA, J.; and NYGODE, B. "A Review of the Commercial Insects Innovative

Research and Technology Development in Africa". The Conservation and Utilization of Commercial Insects. Proceedings of the First International Workshop on the conservation and utilization of commercial insects held in Nairobi, Kenya, 18-21, August 1997. Eds Raina, S.K.; Kioko, E.N.; and Mwanycky, S.W. ICIPE Science Press. 1999: p3-14.

- RASHAD, S. E.; EWEIS, M. A.; and EL-FATAH, A. "Confinement of honey bee colonies during insecticide application". Proceedings of the Third International Conference on Apiculture in Tropical Climates. Nairobi, Kenya. 5-9 November 1984. IBRA. 1985: p145-148.

- RASHAD, S. E.; EWEIS, M. A.; and NOUR, M. E. "Studies on the infestation of honey bees (Apis mellifera) by Acarapis woodi in Egypt". Proceedings of the Third International Conference on Apiculture in Tropical Climates. Nairobi, Kenya. 5-9 November 1984. IBRA. 1985: p152-155.

- RAVISHANKAR, J. "In Defence of Cement". B/D Journal. Issue No. 79 (June 2006): p3.

- RINDERER, T. "Morphological and Adaptive Variation in Apis mellifera". The Conservation and Utilization of Commercial Insects. Proceedings of the First International Workshop on the conservation and utilization of commercial insects held in Nairobi, Kenya, 18-21, August 1997. Eds Raina, S.K.; Kioko, E.N.; and Mwanycky, S.W. ICIPE Science Press. 1999: p71-73.

- RUTTNER, F. and KAUHAUSEN, D. "Honey bees of Tropical Africa: Ecological Diversification and Isolation". Proceedings of the Third International Conference on Apiculture in Tropical Climates. Nairobi, Kenya. 5-9 November 1984. IBRA. 1985: p45-51.

- SANFORD, M. T. (2009). Diseases and Pests of the Honey Bee. University of Florida. http://edis.ifas.ufl.edu/(copyright.htm)

- SHAH, F. A. "Varroa destroys Apis mellifera: Beekeeping in Kashmir Valley". B/D Journal. Issue No. 79 (June 2006): p3.

- SHIMANUKI, H.; KNOX, D. A.; FURGALA, B.; CARON, D. M.; and WILLIAMS, J. L. "Diseases and Pests of Honey Bees". The Hive and the Honey Bee. Dadant & Sons, Inc. Hamilton, Illinois. 1992: p1083-1151.

- TINGEK, S.; KOENIGER, G.; and KOENIGER, N. "Newly recorded Parasitic Fly of Honey Bees in Sabah, Malaysia". B/D Journal. Issue No. 92 (September 2009): p3-4.

- WHITCOMBE, R. P. "Aspects of the biology and management of Apis florea in Oman". Proceedings of the Third International Conference on Apiculture in Tropical Climates. Nairobi, Kenya. 5-9 November 1984. IBRA. 1985: p96-100.

- WOYKE, J. "Tropilaelaps clareae in Afghanistan, and control methods applicable in Tropical Asia". Proceedings of the Third International Conference on Apiculture in Tropical Climates. Nairobi, Kenya. 5-9 November 1984. IBRA. 1985: p163-166.

- XIANSHU, L. "Advancing Chinese Apiculture". Proceedings of the Third International Conference on Apiculture in Tropical Climates. Nairobi, Kenya. 5-9 November 1984. IBRA. 1985: p93-95.